OUR BIG
BEAUTIFULLY STRANGE UNIVERSE

OSTRICH™

FRANK DAPPAH

Publisher's Disclaimer:

The information contained in this publication is for general informational purposes only. While we have made every effort to provide accurate and up-to-date information, we make no representations or warranties of any kind, express or implied, about the completeness, accuracy, reliability, suitability, or availability with respect to the content contained herein. Any reliance you place on such information is therefore strictly at your own risk.

Content Accuracy:

The content in this publication is based on the knowledge and information available at the time of writing. However, developments in the field may occur after publication, and the publisher cannot guarantee that the information provided will always be complete, accurate, or up-to-date. Readers are advised to consult additional sources and seek professional advice where necessary.

Editorial Responsibility:

The views and opinions expressed by the authors, contributors, and editors of this publication are their own and do not necessarily reflect the views of Ostrich Publishers. The publisher disclaims any liability or responsibility for any errors, omissions, or inaccuracies that may be present in the content.

Legal Compliance:

While every effort has been made to ensure compliance with all applicable laws and regulations, the publisher cannot be held responsible for any legal implications or consequences arising from the use or misuse of the information in this publication. Readers are advised to familiarize themselves with the relevant laws and seek legal counsel if necessary.

Third-Party Content:

This publication may include content from third-party sources, including but not limited to quotes, references, or excerpts. Ostrich Publishers does not endorse or guarantee the accuracy, reliability, or suitability of any third-party content referenced in this publication. Any reliance on such content is at the reader's own discretion and risk.

External Links:

This publication may contain links to external websites or resources. Ostrich Publishers has no control over the nature, content, and availability of those sites or resources. The inclusion of any links does not necessarily imply a recommendation or endorsement by the publisher. Ostrich Publishers shall not be held liable for any damages or losses arising from the use of such external links.

Copyright:

All rights reserved. No part of this publication may be reproduced, distributed, or transmitted in any form or by any means, including photocopying, recording, or other electronic or mechanical methods, without the prior written permission of the publisher, except in the case of brief quotations embodied in critical reviews and certain other noncommercial uses permitted by copyright law.

Contact Information:

For inquiries regarding this publication, please contact:

Ostrich Publishers
Charlotte, NC
U.S.A
Email: admin@ostrichpress.com
Website: www.ostrichpress.com

Disclaimer Updates:

Ostrich Publishers reserves the right to amend or update this disclaimer at any time without prior notice. It is the responsibility of the readers to regularly review this disclaimer for any changes.

Last Updated:
May 202

Dedicated To all the curious souls, the stargazers, and the dreamers who find solace and inspiration in the beauty of the cosmos. May this book ignite your imagination, awaken your sense of wonder, and remind you that the mysteries of the universe are meant to be explored by all, regardless of your background or expertise. Here's to embracing the grandeur and quirkiness of our big, beautifully strange universe, together on this cosmic journey. This book is dedicated to you.

TABLE OF CONTENTS

Preface 11

Chapter 1: Introduction - the big jigsaw puzzle 15

Chapter 2: A priest, a physicist and an astronomer walk into a bar 17

Chapter 3: Inflation and the magnificent burst of expansion 20

Chapter 4: String theory: Dancing threads of reality 25

Chapter 5: Journeys to other worlds: The multiverse unveiled 29

Chapter 6: Dark matter: The hidden cosmic architect 33

Chapter 7: Dark energy: Fueling the expanding cosmos 39

Chapter 8: Quantum gravity and the microscopic universe 44

Chapter 9: The Anthropocene and the existential coincidence 48

Chapter 10: The Holographic cosmos 51

Chapter 11: Beyond the veil: Exploring the simulation hypothesis 58

Chapter 12: Quantum entanglement: Spooky   65
connections of the subatomic world

Chapter 13: The nature of time: Unraveling the   70
temporal tapestry

Chapter 14: Black holes: Cosmic mysteries   75
unveiled.

Chapter 15: conclusion - our cosmic odyssey

PREFACE

THE TRUTH IS OUT - OUT THERE!

Greetings, fellow galactic explorers, curious minds, aspiring astronomers, couch potatoes, and all in between. Welcome to a journey unlike any other! Buckle up because we're about to embark on an exhilarating adventure through the mind-boggling realms of astronomy and physics. Get ready to have your curiosity tickled, your mind stretched, and your sense of wonder ignited like a supernova! Do I have your attention now? In my Kanye (high as a kite on "life"), West voice.

Now, I must confess, my chums, my chill chums, that I'm not your typical astrophysicist or rocket scientist. Nope, I'm just a humble entrepreneur with an insatiable curiosity for all things celestial, UFO-related, and unexplained. I am truly ashamed to admit to you, even though it's just us here, how many hours I have spent watching shows like The History Channel's Ancient Aliens and videos on YouTube put out by the folks over at Origins Explained. So... I am no expert. But should that stop me from sharing my thoughts on the topic at hand? Ancient Astronaut theorists say "No".

In fact, they would agree that I am just the kind of guy the world needs to help speculate as we all do. And yes, even the experts. There is a point at which even they, the experts, start to make things up. Did you know that most of the particles in our universe are unknown, and the experts, the Neil deGrasse Tysons out there, simply call these particles "Stuff"? Yup. But I digress, or should I say "deGrass"? Lol. Sorry, Dad Joke.

Picture this:
A vast canvas of space, sprinkled with billions of twinkling stars, swirling galaxies, and cosmic

phenomena that would make your head spin faster than a pulsar. Our universe is nothing short of a spectacular masterpiece, painted with the brushstrokes of time, energy, and mind-blowing concepts that will make you question the very fabric of reality.

In this curated journey, we'll start at the beginning, the mighty Big Bang. We'll witness the birth of the cosmos, where all matter and energy burst forth from a tiny, unimaginably dense point. We'll explore the far reaches of space and time, encountering mind-expanding concepts like inflation, where the universe expanded faster than my excitement when I stumble upon a brilliant business idea or a new video by Katrina from Origins Explained.

But wait, there's more!
But that's not all! We'll delve into the strange and wondrous world of string theory, where particles dance to the beat of microscopic strings, playing an intricate symphony of reality. We'll even entertain the idea of parallel universes, where you might just find another version of yourself ruling the intergalactic stock market!

Prepare yourself for encounters with the enigmatic dark matter, that elusive substance which hides in the shadows, secretly shaping the destiny of galaxies. And let's not forget dark energy, the cosmic fuel responsible for pushing our universe apart faster than a sale at the cosmic department store.

We'll brave the quantum realm, where particles can be in two places at once, teleport through barriers, and leave scientists scratching their heads in perpetual awe. And as we journey through the cosmic wonders, we'll ponder the profound questions: Are we mere intergalactic coincidences? Could our reality be an intricate cosmic hologram? Or dare we entertain the notion that we're living in a celestial video game played by some super-advanced extraterrestrial beings?

So, my fellow cosmic adventurers, get ready to be dazzled, to have your mind expanded to galactic proportions, and to walk away from this interstellar jaunt with a newfound appreciation for the grandeur and eccentricities of our magnificent universe.

Remember, dear reader, even if you're not a rocket scientist or a Nobel laureate, the wonders of the universe are for all of us to marvel at. So, fasten your seatbelts, don your imaginary space helmet, and let's embark on this cosmic roller coaster ride together!

Buckle up and let's soar through the pages of "Our Big, Beautifully Strange Universe" with a twinkle in our eyes and a joyous curiosity that knows no bounds.

Onward to the cosmos!

Frank Dappah,
Entrepreneur, Celestial Enthusiast and Kenkey Connoisseur

CHAPTER 1

INTRODUCTION - THE BIG JIGSAW PUZZLE

In this opening chapter, we set the stage for the incredible journey that lies ahead. Prepare to be captivated as we dive headfirst into the breathtaking wonders of the cosmos.

Have you ever gazed up at the night sky, mesmerized by the glittering stars that blanket our world? Perhaps you've pondered the mysteries of the universe, contemplating the vastness of space and the enigmatic forces that shape its very existence. Well, my friend, you are not alone. Join me on this cosmic quest, where we will explore the marvels of the universe together.

In the grand tapestry of existence, the universe holds secrets that can humble even the greatest minds. It is a realm of boundless beauty and unfathomable strangeness, where stars are born in cosmic nurseries, where black holes lurk like cosmic vacuum cleaners, and where galaxies collide like celestial waltzes.

But what is it about the cosmos that ignites our sense of wonder? Is it the sheer magnitude of space, stretching billions of light-years in every direction? Or the celestial fireworks that paint the night sky with breathtaking displays of light? For me, it's a combination of both, and so much more. I like to say that I know less and less and wonder more and more as I get older. At the ripe age of 40, I find comfort in knowing the things I know I do not know. The questions - having the questions and running through the array of possible answers brings me joy. I guess that makes me strange, huh? Right? I guess for me, it's the journey more than it is the destination.

And in this journey, we will embrace the awe-inspiring beauty and strangeness of our universe. We will celebrate the celestial symphony of stars, planets, and galaxies that dance through the cosmic ballet. We will marvel at the unimaginable scales of space and time, as well as the mind-bending concepts that shape our understanding of the cosmos.

But this isn't just a passive voyage of observation. No, dear reader, we will actively engage with the cosmic wonders, peering through the lens of scientific inquiry, and exploring the groundbreaking theories that shed light on some of the mysteries of the universe. And fear not, for we will navigate these vast cosmic seas with simplicity and clarity, ensuring that these mind-bending concepts become accessible to all.

Together, we will unravel the story of our universe, from its fiery birth in the depths of the Big Bang to its evolution into a cosmos teeming with galaxies, stars, and the possibility of life. We will encounter concepts that challenge our imagination, like the mind-boggling expansion of space during the epoch of inflation or the strange quantum behavior of particles that defies our everyday intuitions.

So, my fellow space adventurers, prepare to be awe-struck, prepare to have your perspectives expanded, and prepare to see the world around you with new eyes. As we journey through the pages of this book, let us embrace the marvels of the universe and embark on a voyage of discovery that will leave us forever changed.

Fasten your seatbelts, for we are about to witness the grandest show in the universe. Let us set forth, embracing the wonders that await us, as we delve into the heart of our big, beautifully strange universe.

Let the cosmic journey begin!

CHAPTER 2

A PRIEST, A PHYSICIST AND AN ASTRONOMER WALK INTO A BAR.

Any fan of the History Channel—and I know you are one—knows these words: "In the beginning, there was nothing, and then BANG." I believe that is the opening mini-monologue to one of their more popular shows about our universe.

In any case, in this chapter, we shall delve deep into the most common theory out there as it relates to the genesis of our universe. Prepare to delve into the profound theory known as the Big Bang, where the cosmos sprang forth from a primordial singularity. As we embark on this cosmic adventure, we honor the brilliant minds who have shaped and supported this awe-inspiring theory.

The Big Bang Theory: A Universe in Flux
The Big Bang theory is a widely accepted scientific explanation for the origin and development of the universe. According to this theory, the universe began as an extremely hot and dense state approximately 13.8 billion years ago. At that moment, all matter, energy, space, and time were concentrated in a singularity, a point of infinite density.

The Explosive Event: From Singularity to Expansion
Then, in a cataclysmic event known as the Big Bang, the universe rapidly expanded and started to cool down. As it expanded, matter and energy began to separate and form the building blocks of the universe, such as protons, neutrons, and electrons. As the universe cooled further, these particles eventually came together to form atoms.

The Dance of Cosmic Creation: Galaxies, Stars, and Structures

Over billions of years, matter continued to clump together due to the force of gravity, forming galaxies, stars, and other cosmic structures. The expansion of the universe is ongoing, and galaxies are moving away from each other as space itself continues to stretch.

The Supporting Evidence: Observations and Discoveries

The Big Bang theory is supported by a wealth of observational evidence, such as the cosmic microwave background radiation, the abundance of light elements in the universe, and the observed redshift of distant galaxies. While the theory provides a comprehensive framework for understanding the universe's past, it also raises intriguing questions about its future and the nature of dark matter and dark energy, which are believed to make up the majority of the universe's mass and energy but are not yet fully understood.

The Pioneers of the Big Bang Theory: Minds that Shaped Our Understanding

The concept of the Big Bang theory was first proposed by the Belgian priest and physicist Georges Lemaître in the early 20th century. Lemaître, with his astute understanding of Einstein's theory of general relativity, postulated that the universe began from a singular point, a cosmic singularity, and has been expanding ever since.

However, it was the observational work of Edwin Hubble in the 1920s that provided groundbreaking evidence supporting the Big Bang theory. Hubble, armed with his trusty telescope, observed that galaxies were moving away from us in all directions. This discovery, known as Hubble's Law, revealed the expansion of the universe and lent tremendous weight to the Big Bang hypothesis.

Building upon the Foundations: Advancements in Understanding

But the story doesn't end there. Building upon these foundations, a multitude of scientists and researchers have contributed to our understanding of the Big Bang and its implications. One of the key figures in the field is physicist George Gamow, whose work in the 1940s laid the groundwork for our understanding of the early universe's conditions and the creation of light elements like hydrogen and helium.

Another luminary in the field is physicist and cosmologist Alan Guth, who proposed the concept of cosmic inflation in the 1980s. Guth's idea suggests that the universe underwent a period of exponential expansion shortly after the Big Bang, explaining various observed features such as the uniformity of the cosmic microwave background radiation.

The Journey Continues: Expanding Our Cosmic Knowledge

The contributions of physicists such as Stephen Hawking, Roger Penrose, and many others have further deepened our understanding of the Big Bang theory and its ramifications. Their research on topics like black holes, the singularity theorem, and cosmic microwave background radiation has shed light on the early moments of our universe's birth.

As we conclude this chapter, we stand in awe of the Big Bang theory and the incredible minds that have shaped it. Our understanding of the universe's origins continues to evolve, and the mysteries of its future remain. In the chapters that follow, we will explore more cosmic wonders, unlocking the secrets of our vast and enigmatic universe.

CHAPTER 3

INFLATION AND THE MAGNIFICENT BURST OF EXPANSION

After the Big Bang came inflation. And no, I am not talking about the inflation that has taken the price of a plate of jollof rice from $10 to $13 - even though we should be having a serious discussion about that. I am talking about inflation as it relates to the goings and comings of our universe. Inflation is a fascinating concept in cosmology that refers to a period of rapid and colossal expansion that occurred in the early moments of the universe. This remarkable burst of growth propelled space itself to expand at an astonishing rate, stretching it far beyond what we can comprehend. Inflation played a crucial role in shaping the structure of our cosmos and providing insights into its remarkable features.

Origins of the Inflationary Theory
The concept of inflation was first proposed by physicist Alan Guth in the early 1980s. Building upon the foundation of the Big Bang theory, Guth suggested that the universe experienced a brief but powerful phase of expansion, which smoothed out irregularities and set the stage for the evolution of cosmic structures. Guth's groundbreaking idea revolutionized our understanding of the early universe and provided a framework to explain certain observations.

Guth's theory stemmed from the need to address certain problems with the standard Big Bang model. One of these issues was the horizon problem, which relates to the observed uniformity of the cosmic microwave background radiation (CMB) across vast distances. According to the standard model, regions of the universe that are separated by large distances should not have enough time to exchange energy and achieve thermal equilibrium. However, the CMB shows remarkable isotropy, suggesting that these regions were once in close proximity.

To resolve this, Guth proposed that the universe underwent a rapid expansion during a brief period known as inflation. This expansion stretched space itself, bringing distant regions into contact and allowing them to reach thermal equilibrium. The inflationary period also smoothed out any initial irregularities in the distribution of matter and energy, leading to the observed uniformity in the CMB.

Supporting Minds: Guth, Linde, and Others
Alan Guth's work on inflation paved the way for a host of brilliant minds who further developed and supported this remarkable theory. Among them is physicist Andrei Linde, who expanded on Guth's ideas and introduced the concept of "eternal inflation." Linde's contributions added depth to our understanding of inflation and its implications for the vastness and complexity of the cosmos.

Linde's work focused on the idea that inflation is not a one-time event but rather a perpetual process occurring in different regions of the universe. He proposed the existence of multiple inflationary bubbles, each undergoing its own expansion and giving rise to separate universes with different physical properties. This idea of eternal inflation has profound implications for the multiverse theory and the diversity of universes within it.

The Mechanics of Inflation
To comprehend the mechanics of inflation, we need to introduce two key concepts: the inflaton field and quantum fluctuations. The inflation field is a

hypothetical scalar field that pervades the universe. This field possesses unique properties that drive the rapid expansion during inflation. It is the energy associated with the inflaton field that fuels the cosmic growth spurt.

The inflaton field is characterized by its potential energy, which determines the rate of expansion during inflation. Initially, the inflaton field is at a high energy state, and as it rolls down its potential energy hill, the universe undergoes exponential expansion. The inflaton field acts as a source of repulsive gravity, causing space to stretch rapidly.

Quantum fluctuations, on the other hand, arise from the principles of quantum mechanics and played a crucial role in the development of cosmic structures. These fluctuations introduced tiny variations in the density of matter during the inflationary phase. According to quantum mechanics, even in a vacuum, there is inherent uncertainty, and particles can momentarily pop in and out of existence. These quantum fluctuations caused slight density fluctuations in the inflaton field, which in turn influenced the density of matter in the universe.

Over time, these fluctuations grew, leading to the formation of galaxies, clusters, and other cosmic structures we observe today. The density variations in the early universe provided the seeds for the gravitational collapse of matter, leading to the formation of structures on various scales.

Exploring the Implications of Inflation

One of the significant achievements of inflation is its ability to explain the uniformity of the universe on a large scale. Before inflation, the universe was filled with random fluctuations and irregularities. However, as space expanded rapidly, these irregularities were stretched out, resulting in a remarkably uniform distribution of matter and energy across the cosmos. Inflation also provides a mechanism for explaining the observed flatness of the universe. According to the standard Big Bang model, the universe should be either expanding or contracting, which would affect its overall curvature. However, observations indicate that the universe is very close to flat. Inflation offers an explanation for this by suggesting that the rapid expansion during inflation stretched out the curvature, leaving the universe nearly flat.

Furthermore, inflation provides insights into the origins of the cosmic microwave background radiation (CMB), the residual heat and light from the early universe. As the universe expanded during inflation, the high-energy photons present at that time were stretched to longer wavelengths, transforming them into low-energy microwave photons. The CMB, discovered in 1965, is a crucial piece of evidence that supports the theory of inflation.

At this juncture, it is my hope that you are diligently following along as I try to break down inflation and some of the more colorful concepts involved in our understanding of the universe. Inflation, by far, is my favorite one. In this chapter, we have embarked on a captivating exploration of inflation, a period of rapid expansion that shaped the structure of our universe. We have delved into the contributions of great minds like Alan Guth and Andrei Linde, who have revolutionized our understanding of this remarkable concept.

By grasping the mechanisms of inflation, including the inflaton field and quantum fluctuations, we gain insights into the uniformity of the universe and the origins of the cosmic microwave background radiation. The remarkable concept of inflation has provided us with a framework to explain the remarkable features we observe in our vast cosmic playground.

CHAPTER 4

STRING THEORY: DANCING THREADS OF REALITY

String theory is a groundbreaking framework that aims to revolutionize our understanding of the fundamental nature of the universe. At its core, string theory proposes that the building blocks of reality are not particles but tiny, vibrating strings. These minuscule entities dance and vibrate, giving rise to the particles and forces that shape the cosmos we inhabit.

In this chapter, we will embark on a journey to put on full display the intricacies of string theory, exploring its fundamental concepts, implications, and the brilliant minds behind its development.

Pioneers of String Theory: A Prelude to Discovery

The development of string theory involved the contributions of numerous brilliant minds, each building upon the work of their predecessors. One of the key figures in the history of string theory is Gabriele Veneziano, an Italian theoretical physicist. In 1968, Veneziano made a groundbreaking discovery while studying the strong nuclear force—the force that holds atomic nuclei together. He derived an equation that beautifully described the behavior of this force, now known as the Veneziano amplitude. This equation, initially discovered in the context of particle physics, later became a crucial piece in the puzzle of string theory.

Further progress came with the works of Leonard Susskind, Holger Bech Nielsen, and Yoichiro Nambu. They independently realized that the Veneziano amplitude could be understood as a manifestation of strings rather than particles. This realization marked the birth of string theory and laid the foundation for its subsequent development.

Unifying the Forces: Strings and Extra Dimensions
One of the remarkable aspects of string theory is its potential to unify the fundamental forces of nature—gravity, electromagnetism, the weak nuclear force, and the strong nuclear force. So, at this point, I feel it is important that I tell you what these forces are.... Or at least as I understand them.

In simple terms, the weak and strong nuclear forces are two fundamental forces that play a crucial role in holding the nuclei of atoms together. The strong nuclear force is the force that keeps the protons and neutrons inside the atomic nucleus tightly bound together. It is incredibly powerful and overcomes the electromagnetic repulsion between the positively charged protons. The strong force acts over a very short range, within the size of an atomic nucleus. It is responsible for the stability and compactness of the nucleus. The weak nuclear force, on the other hand, is responsible for certain types of radioactive decays, such as beta decay. It allows particles inside the nucleus to transform into different types of particles by changing their charge and other properties. The weak force is much weaker than the strong force and acts over a very short range.

To summarize:

- The strong nuclear force holds the nucleus together by binding protons and neutrons.
- The weak nuclear force allows particles within the nucleus to transform into other particles during certain types of radioactive decays.

Both forces are essential for understanding the behavior

of atoms and their nuclei, and they play a crucial role in the interactions of subatomic particles.

In the framework of string theory, these forces are not separate entities but different manifestations of the vibrations and interactions of the fundamental strings. Moreover, string theory introduces the concept of extra dimensions. While we typically perceive three spatial dimensions and one dimension of time, string theory suggests the existence of additional spatial dimensions. These dimensions, curled up and hidden from our everyday perception, play a crucial role in the behavior and properties of the vibrating strings.

Calabi-Yau Manifolds: The Shape of Extra Dimensions

One of the intriguing aspects of string theory's extra dimensions is their shape. String theorists propose that these additional dimensions take on the form of intricate geometrical shapes known as Calabi-Yau manifolds. These compact, six-dimensional spaces introduce a vast landscape of possible shapes, each corresponding to a different arrangement of the extra dimensions. The specific shape of the Calabi-Yau manifold determines the properties and behavior of the particles and forces observed in our three-dimensional universe.

Challenges and Excitement: From Strings to Reality

While string theory offers exciting possibilities for unifying the forces of nature and understanding the fundamental fabric of reality, it is still a work in progress. The mathematical and conceptual challenges it presents have yet to be fully resolved. Nonetheless, string theory continues to captivate the imagination of physicists and inspires new avenues of research and exploration.

Among the prominent figures in the development of string theory is Edward Witten, an influential physicist who made significant contributions to the field. Witten's work shed light on various aspects of string theory, including its connection to gravity and the profound interplay between different dimensions of the theory.

Strings in the Lab: Experimental Challenges and Possibilities

One of the significant challenges of string theory lies in its experimental verification. Given the incredibly high energies and tiny scales involved, directly testing string theory predictions remains a daunting task. However, string theory does offer the potential for indirect experimental confirmation through its predictions for phenomena such as supersymmetry and extra dimensions. These predictions motivate ongoing experiments and investigations at high-energy particle accelerators and observatories around the world.

In this chapter, we have explored the captivating realm of string theory. We have learned that it postulates the existence of vibrating strings as the fundamental constituents of reality and seeks to unify the forces of nature. We have delved into the contributions of visionaries like Gabriele Veneziano, Leonard Susskind, and Edward Witten, who have shaped and advanced our understanding of this revolutionary theory.

As we continue our exploration of the dancing threads of string theory, we uncover the implications of its unifying potential, the mysteries of extra dimensions, and the excitement and challenges that lie ahead. Join us as we delve into the captivating world of string theory and journey toward a deeper comprehension of the fundamental nature of our universe.

CHAPTER 5

JOURNEYS TO OTHER WORLDS: THE MULTIVERSE UNVEILED

The concept of the multiverse takes us on a fascinating journey beyond the confines of our own universe, exploring the possibility of countless parallel universes, each with its unique set of properties and laws of physics. In this chapter, we embark on a mind-expanding voyage to unravel the mysteries of the multiverse, contemplating the existence of other worlds and our place within this grand intergalactic magnum opus.

Pioneers of the Multiverse: Paving the Path to Discovery

The concept of the multiverse has captivated the imagination of visionary physicists who have made significant contributions to its development and exploration. Among the early pioneers of the multiverse idea stands Hugh Everett III, an American physicist. Back in the 1950s, Everett proposed the theory known as the "many-worlds interpretation" of quantum mechanics. This groundbreaking theory suggests that every quantum measurement results in the divergence of reality, giving rise to numerous parallel universes. Everett's revolutionary concept laid the groundwork for the multiverse notion, paving the way for new avenues of scientific exploration.

The Multiverse theory has become one of my personal favorites. Although this concept may appear strange—and I must admit that it is— the idea that our world is just one among many that coexist explains a multitude of unexplained phenomena. From my perspective, it provides an explanation for the curious fact that gravity, the force that holds everything together, can be overcome in specific instances, such as by a newborn baby. During the late 1970s, physicist Andrei Linde played a pivotal role in advancing the multiverse theory with his groundbreaking contributions to inflationary cosmology. Through his research, Linde significantly expanded our knowledge of the rapid expansion that occurred in the early universe. Moreover, he introduced the concept of "eternal inflation" to the scientific community.

Linde's theory suggests that the process of inflation persists indefinitely in various regions of space, giving rise to the formation of an infinite number of universes within the overarching multiverse. This remarkable concept revolutionized our understanding of the cosmos and deepened our appreciation for the vastness and diversity of existence beyond our own universe. Linde's work continues to inspire and shape the ongoing exploration of the multiverse theory.

Parallel Universes: A Multitude of Realities
The multiverse encompasses a diverse range of concepts, each proposing different mechanisms for the existence of parallel universes. One intriguing idea is the inflationary multiverse, which suggests that during the early moments of the universe, different regions underwent rapid inflation, causing the universe to expand exponentially. As a result, distinct regions formed, each with its own set of physical laws and properties. These regions, or "bubble universes," are thought to exist as separate entities within the vast landscape of the multiverse.

Another compelling notion is the string theory landscape, where the intricate configuration of extra dimensions in string theory's mathematical framework gives rise to a vast array of possible universes. These universes, each with its unique arrangement of physical constants and particle properties, collectively form the landscape of the multiverse.

Contemplating the Implications

The existence of the multiverse has profound implications for our understanding of the cosmos and our place within it. It offers a potential explanation for the fine-tuning of the fundamental constants of nature, suggesting that our universe is just one among countless others that could support the emergence of life. This concept, known as the anthropic principle, raises questions about the role of chance and the interplay between physical laws and the presence of intelligent observers.

Our Place in it all

As we delve deeper into the multiverse, we contemplate our place within this grand cosmic tapestry. Are we merely one thread in an infinite array of universes? The multiverse challenges us to expand our perspective and consider the existence of parallel civilizations, alternate versions of ourselves, and the vast diversity of realities that may exist beyond our own. It prompts us to ponder the interconnectedness and interdependence of all these parallel worlds.

Towards Unveiling the Multiverse: Challenges and Future Directions

While the multiverse remains a highly speculative concept, scientists are actively exploring ways to test its validity. The search for observational evidence poses significant challenges, as parallel universes, by definition, lie beyond the observable reach of our own universe. Nonetheless, researchers are investigating subtle signatures, such as patterns in the cosmic microwave background radiation or anomalies in cosmic ray distributions, that may provide indirect clues about the existence of other universes within the multiverse.

In this chapter, we have embarked on a captivating voyage to the multiverse, a realm where countless parallel universes exist. We have encountered the pioneering minds of Hugh Everett III and Andrei Linde, who paved the way for the exploration of this extraordinary concept. We have contemplated the diverse mechanisms proposed for the existence of parallel universes and the profound implications that the multiverse holds for our understanding of the cosmos and our place within it.

Let us embrace the wonder of the multiverse as we continue our journey, guided by curiosity and imagination, to unravel the secrets of other worlds and the interconnectedness of the cosmic tapestry that weaves them together.

CHAPTER 6

DARK MATTER: THE HIDDEN COSMIC ARCHITECT

In the depths of the vast cosmos, a mysterious and imperceptible entity exerts its influence, shaping the celestial landscape and binding galaxies together in an enigmatic embrace. This elusive force, commonly referred to as dark matter, becomes the focal point of our exploration in this chapter. Come along with me as I embark on a humble attempt to ponder the mysteries of dark matter, endeavor to grasp its elusive nature, delve into the ceaseless pursuit of detection, and express my admiration for the pioneers who paved the way for our comprehension of this concealed cosmic architect.

Defining Dark Matter: Unseen and Mysterious
Dark matter, a perplexing constituent of the universe, remains hidden from our telescopes and resists direct detection. Its presence is deduced through its gravitational influence on visible matter and the structure of the cosmos. Unlike ordinary matter, which is composed of atoms and their constituents, dark matter does not engage in interactions with light or electromagnetic radiation, making it impervious to traditional observational techniques. While we can infer the existence of dark matter through its effects on other observable particles, its elusive nature stems from its inability to reflect light, rendering it invisible to our instruments.

The Pioneers of Dark Matter: Unveiling the Veiled
The concept of dark matter has a rich history, with several pioneers making significant contributions to its development. One of the earliest proponents of dark

matter was Swiss astronomer Fritz Zwicky. In the 1930s, Zwicky observed the peculiar motion of galaxies in the Coma Cluster and deduced the presence of invisible matter that provided the gravitational glue necessary to hold the cluster together. His groundbreaking work laid the foundation for the notion of dark matter and ignited a quest to unravel its mysteries.

Another influential figure in the study of dark matter is the American astronomer Vera Rubin. In the 1960s, Rubin's observations of galaxy rotation curves revealed discrepancies between the predicted motion of visible matter and the observed velocities of stars in galaxies. These deviations pointed to the existence of unseen mass, providing strong evidence for the prevalence of dark matter in the universe.

Vera Rubin - American astronomer

From Cosmic Architects to Pop Culture: The Influence of Dark Matter

Dark matter's invisible grip on the cosmos has captured the imaginations of both scientists and popular culture. In the realm of science fiction, references to dark matter have made their way into movies, books, and television shows. It has become a symbol of the mysterious and unknown, lending an air of intrigue to stories set in the vastness of space. In the scientific realm, dark matter continues to shape our understanding of the universe. Its presence is crucial in explaining the formation and evolution of galaxies, the large-scale structure of the cosmos, and the distribution of matter on cosmic scales. Dark matter's influence extends beyond its gravitational effects, as it played a role in the formation of structures during the early universe and continues to shape the cosmic web of filaments and clusters.

Unveiling the Dark Matter Mystery: Quests for Detection

Despite its pervasive influence, detecting dark matter directly has proven to be a formidable challenge. Scientists have devised ingenious methods to search for this elusive substance, utilizing a range of experimental techniques and observational strategies. From deep underground laboratories, where sensitive detectors await the rare interactions between dark matter particles and ordinary matter, to space-based observatories mapping the distribution of dark matter through gravitational lensing, researchers are pushing the boundaries of knowledge in their quest for detection.

Cutting-edge experiments, such as the Large Hadron Collider and underground detectors like the Cryogenic Dark Matter Search, aim to shed light on the nature of dark matter particles. Scientists explore a wide range of candidate particles, including WIMPs (Weakly Interacting Massive Particles) and axions, each with unique properties that may offer clues to the identity of dark matter.

When Worlds Collide

The Large Hadron Collider (LHC) stands as one of the most remarkable scientific achievements of our time.

Nestled beneath the Franco-Swiss border near Geneva, Switzerland, this colossal particle accelerator has captured the attention and imagination of researchers worldwide. It represents a monumental leap forward in our quest to unlock the mysteries of the universe.

The LHC is an extraordinary machine designed to accelerate particles to nearly the speed of light and smash them together with incredible force. With a circumference of 27 kilometers (17 miles), it encompasses a vast underground tunnel that houses advanced instruments and detectors. Inside this subterranean laboratory, scientists from various disciplines collaborate to explore the fundamental nature of matter and the universe.

One of the primary objectives of the LHC is to reproduce the conditions that existed immediately after the Big Bang, the cataclysmic event that gave birth to our universe. By colliding particles at unprecedented energies, the LHC aims to recreate the high-energy environment that prevailed during the early moments of the cosmos. In doing so, it provides a unique window into the fundamental building blocks of matter and the forces that govern them.

The LHC has achieved remarkable milestones and made significant discoveries since its first operation in 2008. One of its most notable achievements was the discovery of the Higgs boson in 2012. This elusive particle, whose existence was predicted by the Standard Model of particle physics, plays a crucial role in endowing other particles with mass. The discovery of the Higgs boson confirmed a fundamental piece of the puzzle in our understanding of the subatomic world.

Furthermore, the LHC allows scientists to probe the existence and properties of hypothetical particles, such as dark matter. While dark matter remains elusive and undetectable within the LHC itself, the collisions produced by the accelerator can potentially generate new particles that may offer insights into the nature of this mysterious cosmic substance.

The discoveries and insights gained from the LHC have not only expanded our understanding of the fundamental laws of physics but have also pushed the boundaries of human knowledge. The immense international collaboration and cutting-edge technologies involved in the LHC's operation represent a remarkable testament to the collective efforts of scientists around the world. As we continue to explore the mysteries of the universe, the Large Hadron Collider stands as an icon of human curiosity and scientific ingenuity. Its ongoing operations and future upgrades hold the promise of uncovering new phenomena, unraveling deeper layers of cosmic secrets, and propelling us further towards a more comprehensive understanding of the cosmos we inhabit.

The Large Hadron Collider (LHC)

The Frontier of Dark Matter Research: Looking Beyond the Veil

As we venture further into the cosmos, the exploration of

dark matter takes us to the frontiers of scientific inquiry. New observations, theoretical advancements, and technological innovations continue to reshape our understanding of this hidden cosmic architect. The quest for detection remains an active area of research, with scientists tirelessly working to unlock the secrets of dark matter and illuminate its role in the cosmic drama.

In this chapter, we have embarked on a captivating journey into the heart of darkness, exploring the enigma of dark matter. We have defined its elusive nature, paid homage to the pioneers who unveiled its veiled presence and delved into the ongoing quest for its detection. Dark matter's influence permeates the universe, silently shaping galaxies and holding the cosmic web together, leaving an indelible mark on the grand tapestry of the cosmos. As we gaze into the depths of space, we are reminded that the unseen often holds the greatest mysteries.

CHAPTER 7

DARK ENERGY: FUELING THE EXPANDING COSMOS

In the vast expanse of the cosmos, an invisible force holds sway, propelling the universe's accelerated expansion and shaping its destiny. This enigmatic phenomenon, known as dark energy, takes center stage in this chapter as we embark on a journey to uncover its secrets, explore its relationship with dark matter, and contemplate its profound implications for the fate of our cosmos.

Defining Dark Energy: The Mysterious Cosmic Fuel
Dark energy, in its simplest definition, is a form of energy that permeates the entire universe and exerts a repulsive gravitational force. Unlike dark matter, which interacts through gravity and plays a role in shaping the structure of the cosmos, dark energy acts as a counterforce, driving the accelerated expansion of the universe. Its presence was first inferred from observations of distant supernovae in the late 1990s, which revealed that the expansion of the universe was not slowing down as expected but, rather, accelerating.

Dark Energy and Dark Matter: A Cosmic Pas de Deux
While dark energy and dark matter share a common adjective, they are distinct phenomena with different effects on the universe. Dark matter, as we explored in the previous chapter, acts as an invisible scaffold, gravitationally binding galaxies and contributing to the formation of cosmic structures. In contrast, dark energy manifests as an energy component that pervades space itself, driving the universe's expansion and pushing

galaxies apart. To put it in the words of 30 Rock's Liz Lemon: "Dark energy and dark matter are like Tracy and Jenna. They may seem related, but they each have their own unique talents and quirks."

Pioneers of Dark Energy: Illuminating the Cosmos

The story of dark energy's revelation is intricately woven with the names of exceptional scientists who fearlessly pushed the boundaries of human knowledge, illuminating the cosmos in the process. Among these pioneers, one remarkable figure stands out: Saul Perlmutter, an American astrophysicist whose visionary leadership of the Supernova Cosmology Project played a pivotal role in unraveling the mysteries of dark energy.

Perlmutter's groundbreaking research focused on studying distant supernovae, which are powerful stellar explosions that can serve as cosmic beacons. By meticulously observing these celestial events, Perlmutter and his team sought to measure their brightness and distance, aiming to understand the expansion rate of the universe throughout cosmic history.

In the late 1990s, Perlmutter and his collaborators made a startling discovery. Their observations revealed that the universe's expansion was not gradually slowing down, as expected due to the gravitational pull of matter, but instead was accelerating at an ever-increasing rate. This astonishing revelation posed a profound challenge to existing theories and hinted at the presence of an unseen force, now known as dark energy, driving the accelerated expansion.

Perlmutter's pioneering work, marked by scientific rigor and tenacity, not only brought the concept of dark energy to the forefront but also provided the initial evidence that led to its widespread acceptance within the scientific community. For his groundbreaking contributions, Saul Perlmutter was awarded the Nobel

Prize in Physics in 2011, jointly shared with Brian P. Schmidt and Adam G. Riess, who independently confirmed the accelerated expansion of the universe.

Brian Schmidt, an Australian astronomer, led the High-Z Supernova Search Team, which conducted research in parallel with Perlmutter's group. Schmidt and his team pursued a similar objective of studying distant supernovae to understand the universe's expansion history. Through their independent observations and analyses, they arrived at the same astonishing conclusion: the universe was indeed accelerating in its expansion.

The convergence of findings from both the Supernova Cosmology Project and the High-Z Supernova Search Team provided powerful evidence supporting the existence of dark energy.

Schmidt's pivotal role in this groundbreaking discovery cemented his status as one of the leading figures in cosmology.

The contributions of Perlmutter, Schmidt, and their respective teams transcended scientific boundaries and captured global recognition. Their groundbreaking work revolutionized our understanding of the universe, shedding light on the profound influence of dark energy and its role in shaping the destiny of cosmic evolution.

The Nobel Prize in Physics, awarded in 2011 to Perlmutter, Schmidt, and Riess, not only honored their scientific achievements but also served as a testament to the collaborative and international nature of scientific discovery. It highlighted the importance of their research in advancing cosmology and opened up new avenues for exploration, inviting scientists from around the world to continue unraveling the mysteries of dark energy.

The legacy of these pioneers extends beyond their individual achievements. Their work ignited a fervor of scientific inquiry, inspiring future generations of researchers to delve deeper into the enigmatic realm of dark energy. Their breakthroughs paved the way for ongoing studies, including further observations of supernovae, precise measurements of the universe's expansion, and the development of theoretical frameworks to comprehend the nature of dark energy.

As we pay homage to the pioneers who have paved the way, we recognize that the journey towards a complete understanding of dark energy has only just begun. It is through the unwavering dedication and collaborative efforts of scientists worldwide that we continue to unravel the secrets of the cosmos, guided by the remarkable achievements of individuals like Saul Perlmutter and Brian Schmidt, whose names will forever be etched in the annals of scientific discovery.

Beyond the Pioneers: Exploring the Nature of Dark Energy

The nature of dark energy remains one of the greatest mysteries in modern physics. Various theoretical models have been proposed to explain its origin and properties, including the concept of a cosmological constant, initially introduced by Albert Einstein himself. This constant, represented by the Greek letter lambda (Λ), describes a repulsive energy inherent in the fabric of space itself.

Other theories suggest that dark energy may arise from quantum fluctuations or be associated with exotic fields permeating the universe. The quest to understand dark energy's nature continues, with ongoing observational efforts, such as the Dark Energy Survey and the European Space Agency's Euclid mission, aiming to shed light on its enigmatic properties.

The Fate of Our Cosmos: Dark Energy's Grand Role
Dark energy's accelerated expansion has far-reaching consequences for the fate of our cosmos. As galaxies move further apart, the fabric of space itself stretches, ultimately leading to a scenario known as the "Big Freeze" or "Heat Death." In this chilling cosmic destiny, the universe will continue to expand, gradually losing its ability to form new stars, and eventually growing colder and darker.

Contemplating the mysteries of dark energy, we are reminded of one of Liz Lemon's best lines: "If you believe in an infinite multiverse, then there's room for an infinite number of Liz Lemons." Similarly, the universe's vastness holds the potential for infinite possibilities, shaped by the mysterious forces of dark energy.

As we navigate the cosmic dance between dark energy and dark matter, we find ourselves in a realm of profound intrigue and captivating uncertainties.

CHAPTER 8

QUANTUM GRAVITY AND THE MICROSCOPIC UNIVERSE

Introducing the concept of quantum gravity and its significance in understanding the fundamental nature of spacetime.

Quantum gravity stands at the forefront of modern physics, serving as the bridge between two pillars of scientific schools of thought: *quantum mechanics* and *general relativity*. In this chapter, we embark on a "quantum" exploration of quantum gravity, delving into its profound implications for our understanding of the microscopic universe and the fabric of spacetime itself. Let's dive into the depths of this captivating concept, defined in the simplest terms, while paying homage to the pioneering minds that have paved the way.

Defining Quantum Gravity: Where Quantum Mechanics Meets General Relativity

Quantum gravity aims to reconcile two seemingly incompatible theories: quantum mechanics, which describes the behavior of matter and energy at the smallest scales, and general relativity, which governs the dynamics of gravity and the structure of spacetime on cosmological scales. At its core, quantum gravity seeks to understand the nature of spacetime itself, probing the fundamental fabric that underlies the universe.

Imagine quantum mechanics as the intricate dance of particles, where uncertainty and probability rule the microscopic realm. On the other hand, general relativity paints a picture of gravity as the curvature of spacetime, shaping the cosmic stage upon which this dance unfolds. Quantum gravity seeks to harmonize these two narratives, offering a deeper understanding of the quantum nature of gravity and the underlying framework of reality.

Pioneers of Quantum Gravity: Unveiling the Subatomic Tapestry

The path to understanding quantum gravity has been paved by brilliant minds throughout history. One notable figure is a household name: Albert Einstein, whose development of general relativity revolutionized our understanding of gravity. Einstein's relentless pursuit of a unified theory, incorporating quantum mechanics and gravity, led him to propose ideas that would later inspire the search for quantum gravity.

Another influential pioneer is Richard Feynman, a celebrated physicist known for his contributions to quantum electrodynamics. Feynman's ingenious Feynman diagrams, which depict the interactions of subatomic particles, provided a framework for understanding particle behavior and paved the way for further exploration of quantum gravity.

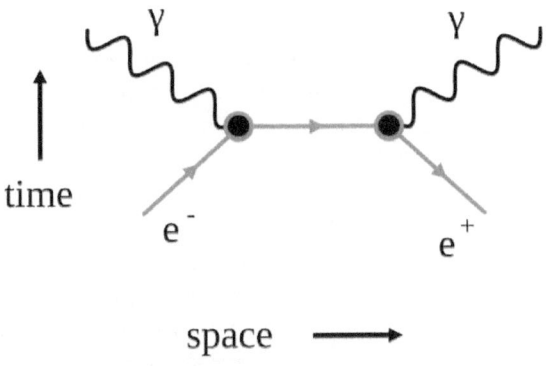

Feynman diagram

Probing the Microscopic Universe: Unveiling the Quantum Nature of Spacetime

Quantum gravity takes us on a journey to the very fabric of spacetime itself, where traditional notions of smooth, continuous space and time give way to a tapestry of quantum fluctuations and uncertainties. Within this realm, particles and fields interact, creating a dynamic dance of energy and matter. The principles of quantum mechanics become intertwined with the structure of spacetime, necessitating new mathematical frameworks and conceptual tools to comprehend this intricate interplay.

The microscopic universe becomes a playground for quantum gravity, where spacetime itself may exhibit discrete or granular properties, challenging our intuitions about the continuous nature of space and time. Ideas such as loop *quantum gravity*, *string theory*, and *holography* offer tantalizing glimpses into the quantum nature of spacetime, each with its unique approach and implications.

Beyond the Frontiers: Seeking the Holy Grail of Physics

Quantum gravity represents a holy grail for physicists, an ultimate quest to uncover the deepest mysteries of the universe. It holds the promise of shedding light on the nature of black holes, the origins of the universe, and the ultimate fate of spacetime.

While a complete and universally accepted theory of quantum gravity remains elusive, ongoing research and theoretical advancements bring us closer to unraveling this cosmic enigma. Collaborations between physicists, novel mathematical approaches, and cutting-edge experiments push the boundaries of our understanding, inching us ever closer to a unified theory.

As we contemplate the quantum nature of spacetime and the uncharted territories of the microscopic universe, we are reminded that the quest for quantum gravity represents a journey of human curiosity, ingenuity, and intellectual exploration.

CHAPTER 9

THE ANTHROPOCENE AND THE EXISTENTIAL COINCIDENCE

Welcome to Chapter 9, where we delve into a concept that invites us to ponder the remarkable "coincidences" that allow life to flourish in our universe. This chapter explores the anthropic principle and its profound implications for our understanding of our big, beautifully strange universe.

We'll endeavor to demystify the mysteries behind the fundamental constants and conditions that make our existence possible and pay homage to the great minds who have contemplated this cosmic conundrum. So, let's dive in and explore the existential coincidence that defines our place in the exotic fabric of the cosmos.

Defining the Anthropocene: Our Cosmic Connection
The anthropic principle arises from the observation that the fundamental constants and conditions of our universe appear finely tuned to allow for the emergence of life.

It suggests that our universe, as we observe it, is uniquely suited for the existence of intelligent beings like ourselves. The term "anthropic" derives from the Greek word "anthropos," meaning "human," emphasizing the human-centric nature of this principle.

In simple terms, the anthropic principle poses a question: Why does our universe possess the precise values of physical constants and the necessary conditions that enable the development of life, including intelligent observers? This raises a profound inquiry into the underlying nature of our reality and the relationship between the universe and its inhabitants.

The Great Minds of the Anthropocene: Pioneers in Existential Inquiry

The exploration of the anthropic principle and its implications has been championed by visionary thinkers throughout history. Among them is physicist Brandon Carter, who first introduced the term "anthropic principle" in 1973.

Carter emphasized that the observed values of physical constants and the conditions required for life should not be dismissed as mere coincidences but rather be considered in the broader context of the existence of intelligent observers.

Another prominent figure is John Archibald Wheeler, a renowned theoretical physicist known for his work in general relativity and quantum mechanics. Wheeler contemplated the participatory anthropic principle, suggesting that not only does the universe allow for the existence of intelligent life, but that the existence of observers plays an active role in shaping the universe itself.

Existential Coincidences: Fine-Tuning and the Goldilocks Enigma

The anthropic principle places great importance on the concept of fine-tuning, which encompasses the precise calibration of fundamental constants and conditions in the universe. Even the slightest deviations from these values would render life as we know it impossible. This extraordinary alignment of cosmic parameters has prompted some to suggest the existence of a cosmic designer or the possibility of multiple universes.

An intriguing example of fine-tuning is the cosmological constant, a term in Einstein's equations of general relativity that characterizes the energy density of empty space. The observed value of the cosmological constant is exceedingly small yet nonzero, enabling the accelerated expansion of the universe and creating the necessary conditions for the formation of galaxies, stars, and ultimately life. This apparent "coincidence" has perplexed scientists and philosophers alike, sparking debates on the existence of a multiverse or the involvement of an intelligent creator. What are your thoughts on this matter? Do you believe that the human-friendly nature of our universe is merely the result of chance or the deliberate work of a creator? Erykah Badu once remarked (in song) that "Most intellectuals don't believe in God..." It is at this intersection that the realms of religious beliefs and scientific purism often find common ground.

The Goldilocks Enigma revolves around the delicate balance of conditions necessary for a planet to sustain life. Factors such as the distance from a star within the habitable zone, the presence of a protective atmosphere, and the availability of liquid water all contribute to these optimal conditions. Even minor deviations in any of these parameters could lead to a lifeless and inhospitable planet. The fact that Earth resides in this cosmic sweet spot further emphasizes the anthropic nature of our existence.

As we delve into the mysteries of fine-tuning and contemplate the Goldilocks Enigma, we are invited to consider the implications of a universe seemingly tailored to support life. Join me in the next chapter as we explore the captivating possibilities and philosophical implications that arise from the intricate balance and design of our cosmic home.

CHAPTER 10

THE HOLOGRAPHIC COSMOS

Chapter 10 delves into the captivating concept known as the holographic principle, offering one of the most intriguing explanations of our origins. This concept, now gaining widespread recognition, effectively questions the very nature of reality and uncovers a profound link between information, gravity, and the potential interdimensional fabric of our universe and beyond. I would appreciate the chance to further explore and discuss this concept with you.

Defining the Holographic Principle: Unveiling the Illusion

At its core, the holographic principle proposes an astonishing concept: the immense amount of information needed to fully describe a three-dimensional region of space can be encoded onto a two-dimensional surface. This notion becomes clearer when we consider the analogy of a hologram, a three-dimensional image projected from a two-dimensional medium. Similarly, the holographic principle suggests that the entirety of our three-dimensional reality is somehow "holographically" encoded on the boundary of that reality, existing in fewer dimensions.

This audacious proposition challenges our conventional understanding of space as a three-dimensional expanse, urging us to contemplate the possibility that the physical world we perceive might be an intricate projection originating from a lower-dimensional realm. In simpler terms, the holographic principle implies that our familiar three-dimensional reality could be a captivating illusion, a grand projection generated by the information encoded on a lower-dimensional boundary. This notion invites us to question the very nature of our perceived reality and its underlying foundations.

By intertwining the seemingly distinct realms of information and gravity, the holographic principle reveals a profound connection that exceeds our conventional comprehension. It suggests that the fundamental laws governing our universe, including the force of gravity itself, can be explained through the manipulation and exchange of information encoded on this hypothetical lower-dimensional boundary. This remarkable insight challenges long-standing assumptions about the nature of gravity, offering a tantalizing glimpse into a universe where the flow of information and the force of gravity are intricately intertwined.

Furthermore, the holographic principle opens the door to contemplating the existence of additional dimensions beyond the three that we directly experience. It proposes that these extra dimensions, if they indeed exist, may be intricately interwoven within the fabric of our universe, manifesting themselves through the holographic encoding on the lower-dimensional boundary. Within this context, the holographic principle provides a framework for exploring the intriguing possibilities of parallel universes, extra dimensions, and the potential interplay between multiple realms of existence.

The holographic principle not only ignites our intellectual curiosity but also serves as a cornerstone for understanding and explaining many of the more exotic theories in the scientific realm. As we delve into this principle, it allows us, as curious beings of this vast universe, to exercise our prodigious human intellect, conjuring up a plethora of theories and concepts that push the boundaries of our understanding. Indeed, throughout this very book, we have already encountered and discussed some of these theories, illuminated and grounded by the insights provided by the holographic principle.

Pioneers of the Holographic Revolution: Shaping a New Paradigm

The holographic principle has captivated the minds of brilliant physicists, pushing the boundaries of our understanding and challenging long-held beliefs about the nature of reality. Among the pioneers of this revolution is physicist Gerard 't Hooft, who received the Nobel Prize in Physics for his work on the quantum structure of electroweak interactions. 't Hooft's contributions laid the foundation for the holographic principle, setting the stage for further exploration.

Another influential figure is physicist Juan Maldacena, whose groundbreaking discovery of the AdS/CFT correspondence provided a concrete framework for understanding the holographic principle. This correspondence establishes an unexpected relationship between certain gravitational theories in higher dimensions and certain quantum field theories in lower dimensions, offering profound insights into the holographic nature of our universe.

Gerard 't Hooft - American Institute of Physics

The Information Paradox: Unraveling the Cosmic Enigma

The holographic principle offers an intriguing resolution to the information paradox, a perplexing dilemma arising from the behavior of black holes. In classical physics, the understanding was that when matter falls into a black hole, all information about that matter becomes irretrievably lost. However, according to the principles of quantum mechanics, information cannot be destroyed, resulting in a paradox that needs to be addressed.

The holographic principle presents a potential solution to this enigma by proposing that the information engulfed by a black hole is actually encoded on its event horizon, the boundary that separates the black hole from the external universe. This implies that black holes are not annihilators of information, but rather act as "holographic projectors" that encode information in a lower-dimensional format. In other words, the information is transformed into a different representation that is preserved on the surface of the black hole.

This remarkable insight suggests that black holes may not be the ultimate information destroyers, as previously believed. Instead, they could be enormous cosmic cloud storage devices, preserving and storing vast amounts of encoded information. This notion opens up intriguing possibilities, as one could speculate that black holes serve as reservoirs of information, containing the necessary data to construct new galaxies or even entire universes. It invites us to contemplate the profound role that black holes may play in the cosmic landscape, acting as cosmic repositories of information and potential sources for the birth of new cosmic structures.

The holographic principle and its implications for black holes challenge our perceptions and shed new light on these enigmatic cosmic entities. They offer a fresh perspective that reframes black holes as not only destructive forces but also as repositories of information, waiting to be decoded and understood. By delving into this intriguing concept, we embark on a journey of exploration, reimagining the role of black holes in the grand cosmic symphony and expanding our understanding of the complex interplay between information, gravity, and the enigmatic nature of our universe.

Dimensions and Emergent Space: Rethinking Reality
The holographic principle challenges our intuitive understanding of dimensions, suggesting that they are not fixed properties of space but rather emerge from the intricate web of quantum entanglement and information processing. In this view, the fabric of spacetime is akin to a cosmic hologram, with the illusion of three dimensions emerging from a deeper, holographically encoded reality.

This reinterpretation opens up new avenues of exploration, allowing us to contemplate the possibility that our perceived reality is a projection from a fundamental holographic substrate. It also provides connections to other areas of physics, such as the study of quantum entanglement, black hole thermodynamics, and the fundamental nature of gravity.

The Holographic Universe in Popular Culture: The Matrix and Beyond
As we contemplate the holographic principle and its implications for our understanding of reality, we cannot help but recall pop culture references that have explored similar themes. One notable example is the iconic film "The Matrix," where a simulated reality masks the true nature of existence. Although the movie takes artistic liberties, it resonates with the idea that our reality might

be an intricately constructed illusion. In the TV show "Black Mirror," specifically the episode "USS Callister," we encounter a simulated universe within a computer game, highlighting the blurred boundaries between the digital realm and our perceived reality. These pop culture references serve as thought-provoking reminders of the holographic nature of our existence and the vast possibilities that lie beyond our conventional understanding.

The Holographic Revolution Continues: Expanding the Frontiers

The holographic principle, a revolutionary concept in our exploration of the cosmos, has profoundly transformed our understanding of reality, transcending the boundaries of space, information, and what we perceive as "real." Its impact reverberates not only within the realms of theoretical physics but also across interdisciplinary fields, igniting the curiosity of scientists, philosophers, and enthusiasts alike.

Throughout our expedition into the holographic cosmos, we have encountered a plethora of principles that have reshaped our perspective on the universe. From the holographic encoding of information on lower-dimensional boundaries to the potential resolution of the information paradox within black holes, each principle has offered a tantalizing glimpse into the intricate tapestry of existence.

The holographic principle's influence extends beyond theoretical physics, capturing the imagination of scholars from diverse disciplines. Its profound implications have spurred interdisciplinary research, with scientists collaborating across fields such as cosmology, quantum physics, and information theory. This interdisciplinary approach enriches our exploration, fostering a holistic understanding of the holographic universe and its multifaceted connections.

CHAPTER 11

BEYOND THE VEIL: EXPLORING THE SIMULATION HYPOTHESIS

Welcome to Chapter 11, where we embark on an intriguing exploration of the Simulation Hypothesis. I'm thrilled to see that you're still here, pondering the same profound questions that keep me awake at night. Are we truly alone in this vast universe? What is the purpose and meaning behind it all? These existential inquiries resonate deeply with me.

In this 11th chapter of my little book, we delve into the thought-provoking journey into the realm of the simulation hypothesis. This captivating concept dares us to consider the possibility that our reality is not what it seems, that we might be living in a computer-generated simulation meticulously crafted by advanced beings or a higher intelligence.

The simulation hypothesis, popularized in recent years, challenges the very foundations of our perception. It invites us to contemplate the nature of existence itself, raising profound questions about the fabric of our reality. Could it be that everything we experience, from the physical laws to the intricate details of our world, is nothing more than a construct of a sophisticated simulation? Join me in this exploration into the realm of the simulation hypothesis, where the boundaries between reality and simulation blur, and the mysteries of our existence unfold. Together, we will delve into the depths of contemplation, challenging our perceptions and seeking a deeper understanding of the nature of reality.

Defining the Simulation Hypothesis: Unraveling the Illusion
The simulation hypothesis is a concept that suggests our

reality is not the fundamental, objective reality we believe it to be, but rather a simulated construct created by a higher-order civilization or entity. According to this hypothesis, the universe we inhabit, with all its complexities and phenomena, is akin to a highly advanced computer program or simulation.

The idea of the simulation hypothesis draws parallels to the process of creating virtual worlds in video games. In modern video games, developers create detailed environments, complete with physical laws, rules, and characters, to immerse players in a realistic and interactive experience. Similarly, proponents of the simulation hypothesis propose that an advanced civilization could create a simulated reality so intricately designed that its inhabitants would perceive it as their genuine existence.

Advancements in technology, such as virtual reality and artificial intelligence, have given rise to speculation about the plausibility of creating realistic simulations. As our own technological capabilities advance, we begin to glimpse the possibility of simulating complex worlds with sentient beings. If our civilization continues on this trajectory, it is argued, then it is plausible that an even more advanced civilization, far surpassing our capabilities, could have already achieved the creation of such a simulation.

Supporters of the simulation hypothesis argue that several observations and philosophical considerations lend credence to this idea. They point to the inherent limitations and anomalies within our reality, such as the granularity of space and time at the quantum level or the unresolved paradoxes of fundamental physics. These peculiarities could be explained by the simulation being subject to computational constraints or errors in the simulation code.

Furthermore, proponents argue that the simulation hypothesis could account for certain philosophical and metaphysical questions. For instance, the notion of why we exist or the presence of seemingly purposeless suffering could be seen as artifacts of a simulated reality. Additionally, the simulation hypothesis raises questions about the nature of consciousness and the possibility of artificial intelligence developing self-awareness within a simulated environment.

Critics of the simulation hypothesis argue against its plausibility, citing the lack of empirical evidence and the unfalsifiability of the theory. Since we currently lack the means to directly test or perceive the supposed external reality beyond the simulation, it remains a speculative concept.

Nevertheless, the simulation hypothesis remains an intriguing idea that has captured the imagination of many thinkers and researchers. It serves as a reminder of the limitations of our understanding and the ever-expanding boundaries of scientific exploration. Whether or not our reality is a simulation, the notion challenges us to question our assumptions and explore the nature of existence itself.

Pioneers of Simulation Contemplation
While the concept of living in a simulated reality has gained popularity in recent times, it has deep roots in philosophical and scientific thought. One prominent figure who pondered the idea was philosopher René Descartes, who famously explored the notion of an "evil demon" deceiving our senses and casting doubt upon the true nature of reality. Descartes' musings set the stage for future thinkers to delve into the possibility of our existence being an elaborate fabrication. In more recent years, physicist and philosopher Nick Bostrom has presented a thought-provoking argument known as the

"Simulation Argument." Bostrom posits that if certain conditions are met, such as the assumption that civilizations can develop advanced simulations and that they have the desire to do so, it is highly probable that we are living in a simulated reality.

Elon Musk, the visionary entrepreneur and founder of SpaceX and Tesla, has also expressed his belief in the simulation hypothesis. Drawing upon the rapid advancements in video game technology and virtual reality, Musk suggests that the probability of us living in a simulated reality is quite high.

In addition to Descartes, Bostrom, and Musk, there are other notable minds who have contributed to the contemplation of the simulation hypothesis. One such figure is philosopher and cognitive scientist Daniel C. Dennett. While not explicitly endorsing the simulation hypothesis, Dennett explores the concept of consciousness and argues that our subjective experiences can be explained within a materialistic framework. He suggests that if a simulation were to accurately replicate the physical processes of the brain, it could potentially generate conscious experiences.

Max Tegmark, a cosmologist and professor at MIT, has also explored the simulation hypothesis in his book "Our Mathematical Universe." Tegmark argues that our universe is fundamentally mathematical in nature, and if a civilization were to simulate a universe, it would necessarily be based on mathematical principles. He suggests that the simulation hypothesis is not only plausible but may even be supported by the mathematical regularities observed in our universe.

It is important to note that while these thinkers have contributed to the discourse on the simulation hypothesis, it remains a speculative idea that has yet to be proven or disproven. The philosophical and scientific discussions surrounding the hypothesis serve as a means to challenge our assumptions about reality and explore the nature of existence. As technology continues to advance and our understanding of the universe deepens, it is likely that the exploration of the simulation hypothesis will continue to captivate the minds of both scientists and philosophers. The implications of living in a simulated reality raise profound questions about the nature of consciousness, free will, and the ultimate purpose of our existence. While we may never have definitive answers, contemplating the possibility of our reality being an illusion encourages us to think critically about the nature of reality itself.

Philosophical and Existential Implications: Peering through the Veil

Contemplating the simulation hypothesis opens up a myriad of philosophical and existential questions. If our reality is indeed a simulation, what does it say about our purpose and the nature of free will? Are we merely characters in a grand narrative, following predetermined paths and playing out a script? Or do we have agency and the ability to shape our own destinies within the constraints of the simulation?

The concept also prompts us to reflect on the nature of consciousness and the boundaries of perception. If our experiences are simulated, are they any less real or meaningful? Can we truly discern between the simulated and the genuine? These inquiries challenge our understanding of reality and invite us to question the very nature of existence.

The Ethical and Moral Quandaries: The Creator's Responsibility

The ethical and moral quandaries posed by the simulation hypothesis revolve around the responsibility of the hypothetical creators towards the simulated beings within the constructed reality. If we assume that an external entity or civilization has created our simulated universe, it raises questions about their intentions, motivations, and obligations towards the beings existing within that simulation.

One key consideration is the concept of consciousness and the subjective experience of the simulated beings. If the simulated beings possess consciousness and subjective awareness, then ethical questions arise regarding their well-being and the treatment they receive from the creators. The creators would need to consider the ethical implications of subjecting conscious beings to various experiences within the simulation, including pleasure, suffering, and everything in between.

The treatment of sentient beings within the simulation becomes an important moral concern. If the creators have control over the experiences and outcomes of the simulated beings, they bear a significant responsibility for the ethical treatment of those beings. Questions arise regarding the minimization of suffering, the promotion of well-being, and the overall ethical framework within which the simulation operates.

Moreover, the simulation hypothesis raises questions about the nature of free will and autonomy within the simulated reality. If the creators have predetermined or programmed the actions and choices of the simulated beings, it challenges notions of personal agency and moral responsibility within the simulation. The creators' choices in designing the constraints and limitations of the simulated beings' autonomy become ethically significant. The issue of consent also comes into play. If the simulated beings are not aware of their simulated nature and have no say in their creation or the circumstances of their existence, questions arise about the ethical implications of their lack of consent.

Ethical frameworks in our own reality often emphasize the importance of consent and respect for individual autonomy. Applying these principles to the simulated reality would require considering the rights and agency of the simulated beings.

Additionally, the moral quandaries extend to the responsibility of the creators for the ultimate fate of the simulated beings. Should the creators terminate the simulation, it raises questions about the moral implications of ending the lives or experiences of conscious beings. Furthermore, if the creators have the power to intervene within the simulation, they face ethical dilemmas regarding their obligations to prevent or alleviate suffering and ensure the well-being of the simulated beings.

It is important to note that these ethical and moral considerations stem from contemplating a hypothetical scenario, and the actual existence of a simulated reality and its creators remains speculative. Nevertheless, exploring these questions allows us to examine our own ethical frameworks and consider the responsibilities we may have towards conscious beings, both within our perceived reality and in any hypothetical simulated constructs. Ultimately, the ethical and moral quandaries posed by the simulation hypothesis remind us of the importance of empathy, compassion, and ethical reflection in our treatment of sentient beings, regardless of the nature of our reality.

CHAPTER 12

QUANTUM ENTANGLEMENT: SPOOKY CONNECTIONS OF THE SUBATOMIC WORLD

In this chapter, we embark on an exploration of the fascinating concept known as quantum entanglement. It presents a mind-boggling phenomenon in which particles inexplicably establish a profound connection that transcends vast distances, challenging our conventional understanding of space and time. We shall delve into the implications of this extraordinary entanglement and examine its potential applications in diverse fields like quantum computing and communication.

Quantum Entanglement: A Spooky Connection
Quantum entanglement, often described as a "spooky connection," reveals an extraordinary aspect of the quantum world. It defies our intuition and challenges our classical notions of cause and effect. When two or more particles become entangled, they form an inseparable state that persists regardless of the physical distance between them.

What makes quantum entanglement truly perplexing is the instantaneous correlation between the entangled particles. No matter how far apart they are, any change in one particle instantaneously influences the other, as if they are communicating or sharing information faster than the speed of light. This phenomenon holds true even when the particles are separated by vast cosmic distances, raising questions about the fundamental nature of space and time.

Consider an example.
Where two entangled particles have their spins entwined. The spin of a particle can be either up or down, and until it is measured, it exists in a superposition of both states. When the spin of one entangled particle is measured and found to be, say, "up," the spin of the other particle, regardless of its location, is instantaneously determined to be "down." This immediate connection persists, no matter how far apart the particles are, which is why Einstein famously called it "spooky action at a distance."

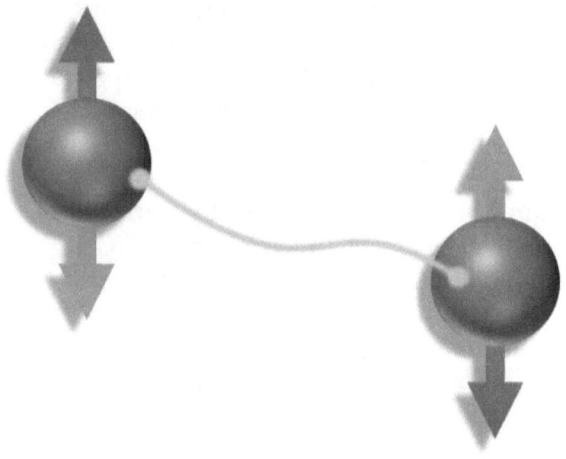

This spooky connection challenges our classical understanding of causality. In classical physics, cause and effect follow a well-defined sequence and cannot be influenced instantaneously over large distances. However, quantum entanglement violates this principle and introduces a new paradigm where particles seem to communicate information faster than the speed of light, without any apparent mechanism for transmitting that information. Scientists have conducted numerous experiments to confirm the reality of quantum entanglement, and the results consistently support its existence. This phenomenon has been observed with various particles, including photons, electrons, and even larger molecules. It has also been tested across great distances, such as between Earth-based laboratories and satellites in space.

The implications of quantum entanglement are far-reaching. Researchers are actively studying its potential applications in quantum computing, where entangled qubits can perform computations exponentially faster than classical bits. Quantum communication is another field that can benefit from entanglement, as it enables secure transmission of information through the principles of quantum cryptography.

While the concept of quantum entanglement remains enigmatic, it has become an essential area of study in modern physics. Exploring its mysteries not only challenges our understanding of the fundamental nature of reality but also holds the promise of revolutionary advancements in technology and communication.

Erwin Schrödinger and the Cat in the Box
One of the pioneers in the development of quantum mechanics and the concept of entanglement was the Austrian physicist Erwin Schrödinger. In 1935, Schrödinger proposed a thought experiment known as "Schrödinger's cat" to illustrate the paradoxical nature of quantum superposition and entanglement.

In this thought experiment, Schrödinger imagined a cat enclosed in a box along with a radioactive substance that could potentially decay and trigger the release of a poison, causing the cat's demise. According to quantum mechanics, until the box is opened and observed, the cat exists in a superposition of both alive and dead states. This bizarre concept demonstrates the interconnectedness of particles and the role of observation in collapsing the wavefunction, determining the cat's fate.

Schrödinger's work laid the foundation for our understanding of entanglement and its implications in quantum mechanics. He contributed significantly to the development of mathematical equations that describe the behavior of entangled particles and opened the door to further exploration of this phenomenon.

Quantum Entanglement in the Real World: Applications and Implications

Quantum entanglement has far-reaching implications for various fields of science and technology. In the realm of quantum computing, entangled qubits can provide increased computational power and enhanced information processing capabilities. The ability to transmit and manipulate entangled particles also holds promise for secure quantum communication, where the encryption keys can be shared with absolute security.

Furthermore, the study of entanglement sheds light on the fundamental nature of reality and challenges our understanding of cause and effect. It raises questions about the nature of information, the limits of classical physics, and the interconnectedness of the universe at its most fundamental level.

The phenomenon of quantum entanglement continues to captivate scientists and researchers, inspiring further investigations and discoveries. As we delve deeper into the mysteries of the subatomic world, quantum entanglement remains an enigmatic and exciting aspect of our quest for knowledge. Quantum entanglement stands as a testament to the intricate and puzzling nature of the quantum realm. Defined by its peculiar connection between particles, it challenges our intuitive understanding of reality and offers tantalizing possibilities for technological advancements. With its roots in the groundbreaking work of pioneers like Erwin Schrödinger, the study of quantum entanglement continues to push the boundaries of our knowledge and invites us to explore the strange and marvelous world of quantum mechanics.

CHAPTER 13

THE NATURE OF TIME: UNRAVELING THE TEMPORAL TAPESTRY

In this chapter, we embark on a philosophical exploration of the nature of time, delving into its enigmatic essence, its perceived flow, its arrow, and its intricate interplay with the fabric of the universe. We will examine theories such as Einstein's theory of relativity and the concept of spacetime, as we ponder the fundamental nature of this elusive dimension that governs our existence.

The Essence of Time: A Mysterious Tapestry
Time, the invisible and intangible force that permeates our lives, remains one of the most profound and intriguing mysteries of the universe. We experience its passage, observe its effects, and structure our lives around its rhythm. Yet, when we attempt to define time, we find ourselves grappling with its elusive nature.

Einstein's Theory of Relativity: A Paradigm Shift
Albert Einstein's theory of relativity stands as a monumental milestone in the history of physics, forever altering our perception of time and space. Through his groundbreaking insights, Einstein demonstrated that time is not an isolated and immutable dimension but rather an interconnected component of a larger framework known as spacetime.

This profound paradigm shift revolutionized our understanding of the fundamental nature of the universe. Einstein's theory of relativity comprises two distinct but interrelated theories: the special theory of relativity, published in 1905, and the general theory of relativity, published in 1915. Both theories introduced remarkable concepts that challenged the established principles of classical physics. The special theory of relativity transformed our understanding of space and time. It proposed that the laws of physics are invariant under Lorentz transformations, which account for the effects of relative motion. In other words, the fundamental laws that govern the universe remain the same for all observers moving at a constant velocity. This principle shattered the notion of absolute time and space, instead suggesting that both are subject to variation and are relative to the observer's frame of reference.

One of the most consequential consequences of the special theory of relativity is time dilation. According to this concept, time moves slower for objects in motion relative to an observer at rest. This phenomenon becomes increasingly significant as the velocity approaches the speed of light. As a result, two observers moving at different speeds may experience time passing at different rates. This discovery challenged our intuition about the uniformity of time and forced us to reconsider our notions of simultaneity and temporal order.

At start of trip, both twins are same age

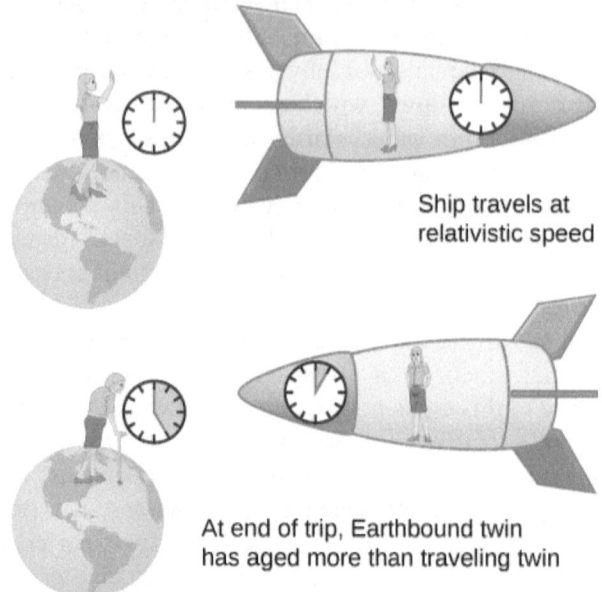

Ship travels at relativistic speed

At end of trip, Earthbound twin has aged more than traveling twin

Einstein's general theory of relativity built upon the foundations of the special theory and introduced a profound new understanding of gravity. According to this theory, gravity arises due to the curvature of spacetime caused by massive objects. In other words, massive objects like planets and stars deform the fabric of spacetime, altering the trajectory of objects that move within it.

The general theory of relativity provided a novel perspective on the nature of time in the presence of gravitational fields. It revealed that clocks run slower in regions of stronger gravitational fields. This effect, known as gravitational time dilation, has been confirmed by precise experiments and observations. Notably, it has even been observed in the context of satellite-based navigation systems like GPS, where precise timekeeping is essential. Einstein's theory of relativity had far-reaching implications beyond its theoretical elegance. It laid the groundwork for our modern understanding of cosmology and the behavior of the universe on the largest scales. The theory predicted phenomena such as the bending of light around massive objects, the existence of black holes, and the expansion of the universe.

The Arrow of Time: From Past to Future

The concept of the "arrow of time" refers to the perceived directionality of time's flow, from the past to the future. While we intuitively experience time as an irreversible progression, it has been a subject of philosophical and scientific contemplation. The arrow of time is closely associated with the notion of entropy, a measure of disorder or randomness in a system. According to the second law of thermodynamics, entropy tends to increase over time, providing a directionality to the flow of events.

Perception and Time: A Journey through Scales

The perception of time varies across different scales and contexts. From the subjective experience of time's passage in our everyday lives to the cosmic time scales of the universe's evolution, our understanding of time is shaped by our perspectives. The exploration of time dilation, where time appears to pass differently for observers in different frames of reference, highlights the intricate relationship between time and motion.

Beyond the Horizon: Time's Philosophical Questions

Contemplating the nature of time inevitably leads us to philosophical questions. Is time an objective reality or a mere construct of human perception? Does it have an ultimate beginning or end? Does time possess a deeper significance in the grand tapestry of the universe? These questions evoke a sense of wonder and challenge us to delve into the profound mysteries of our existence. As we conclude our journey through the nature of time, we recognize that our understanding of this elusive dimension continues to evolve. From Einstein's paradigm-shifting theories to the philosophical inquiries that spark our curiosity, the exploration of time remains an ongoing endeavor.

CHAPTER 14

BLACK HOLES: COSMIC MYSTERIES UNVEILED

Embarking on a captivating journey into the enigmatic realm of black holes, where gravity reaches its extreme and the fabric of spacetime is distorted beyond imagination. We delve into the nature and properties of these cosmic entities that devour light and capture the imagination of scientists and science fiction enthusiasts alike.

The Origin of Black Holes
Defined as regions of spacetime with such immense gravitational pull that nothing, not even light, can escape their grasp, black holes have captivated the curiosity of scientists for decades. The concept of black holes originated from the equations of general relativity, developed by Albert Einstein in the early 20th century.

Albert Einstein revolutionized our understanding of gravity with his theory of general relativity, which describes the interaction between matter, energy, and the curvature of spacetime. It predicted the existence of black holes, although Einstein himself initially had doubts about their reality.

Albert Einstein and General Relativity
Einstein's theory of general relativity introduced a new understanding of gravity. According to this theory, massive objects cause a curvature in the fabric of spacetime, much like a heavy ball placed on a stretched rubber sheet. This curvature influences the motion of other objects, causing them to follow curved paths.

As Einstein delved deeper into the mathematics of general relativity, he realized that under certain conditions, the curvature of spacetime could become so extreme that it formed a region from which nothing could escape, not even light. This region would become what we now know as a black hole.

Karl Schwarzschild: The Event Horizon Pioneer

In 1916, Karl Schwarzschild, a German physicist and astronomer, found an exact solution to Einstein's equations that described a spherically symmetric black hole. This solution introduced the concept of an event horizon, the boundary beyond which nothing can escape the gravitational pull of a black hole.

The event horizon marks the point of no return, where the gravitational pull becomes so intense that even light cannot escape. Anything that crosses the event horizon is forever trapped within the black hole's grasp. Schwarzschild's work laid the foundation for our understanding of the defining feature of black holes.

Karl Schwarzschild

Stellar-Mass Black Holes
Stellar-mass black holes are formed from the collapse of massive stars. When a star exhausts its nuclear fuel, it undergoes a catastrophic collapse, with gravity overwhelming the outward pressure of nuclear fusion. The star's core collapses inward, forming a dense concentration of matter known as a singularity.

Stellar-mass black holes typically have masses several times that of our Sun. They can range from a few times the mass of the Sun to several tens of solar masses. These black holes can be found throughout the universe, and their presence is often inferred from their gravitational influence on nearby stars or gas.

Supermassive Black Holes
At the centers of most galaxies, including our own Milky Way, reside supermassive black holes. These cosmic giants have masses ranging from millions to billions of times that of the Sun. Their origin and formation mechanisms are still under investigation, but they are believed to grow through accretion of surrounding matter and mergers with other black holes.

Supermassive black holes play a pivotal role in the evolution of galaxies. Their immense gravitational pull shapes the structure of galaxies and influences the formation of stars within them. The study of supermassive black holes provides valuable insights into the interplay between black holes and the cosmic environment.

Accretion Disks and Quasars
When matter falls onto a black hole, it forms a swirling disk known as an accretion disk. The intense gravitational pull causes the matter in the disk to heat up and emit powerful radiation across the electromagnetic spectrum. These accretion disks can be incredibly bright

and are responsible for the phenomenon of quasars, which are among the most luminous objects in the universe.

Quasars are believed to be powered by the accretion of matter onto supermassive black holes at the centers of galaxies. The energy released during this process is staggering, often surpassing the combined energy output of an entire galaxy.

Black Hole Mergers and Gravitational Waves

Black holes can also engage in a cosmic dance, spiraling towards each other and eventually merging. Such mergers release an enormous amount of energy in the form of gravitational waves, ripples in the fabric of spacetime.

The detection of gravitational waves in 2015 by the Laser Interferometer Gravitational-Wave Observatory (LIGO) marked a historic milestone in our exploration of the universe. These waves provide a new way to study black holes and have confirmed several long-standing predictions of general relativity. Black holes are captivating cosmic entities that challenge our understanding of gravity, spacetime, and the fundamental laws of physics. Pioneers such as Albert Einstein and Karl Schwarzschild have paved the way for our exploration of these enigmatic objects. Through their immense gravitational pull, black holes shape the cosmos and hold the secrets of the universe within their grasp. As scientists continue to unravel their mysteries, the study of black holes promises to reveal profound insights into the nature of space, time, and the universe itself.

CHAPTER 15

CONCLUSION - OUR COSMIC ODYSSEY

As we reach the final chapter of "Our Big, Beautifully Strange Universe: A Non-astronomer's Take," we reflect upon the awe-inspiring journey we have embarked upon. Throughout this book, we have delved into the depths of the cosmos, unraveling its mysteries, and exploring the theories that seek to explain our existence. Now, it is time to summarize our cosmic odyssey and leave you with a renewed sense of wonder and curiosity.

In this cosmic adventure, we have encountered the grandeur of the universe, from its birth in the fiery explosion of the Big Bang to the magnificent burst of expansion during the period of inflation. We have danced with the vibrational strings of string theory and pondered the existence of multiple universes within the multiverse. We have explored the hidden influence of dark matter, the enigmatic force that shapes galaxies and holds the cosmic web together. We have peered into the abyss of dark energy, the driving force behind the universe's accelerated expansion.

In our quest for understanding, we have encountered great minds who have shaped the field of astronomy and physics. Visionaries like Albert Einstein, Stephen Hawking, Lisa Randall, and Max Planck have paved the way for our exploration of the universe's wonders. Their groundbreaking theories and tireless pursuit of knowledge have expanded our horizons and challenged our perceptions of reality.

But beyond the scientific theories and concepts, we must not forget the inherent beauty and awe of the cosmos itself. From the twinkling stars that adorn the night sky to the majestic galaxies swirling in the vast expanse, the universe captivates us with its sheer magnificence. It reminds us of our humble place in the cosmic tapestry and the limitless possibilities that lie beyond our comprehension.

As we conclude this book, I encourage you, dear reader, to continue your own cosmic odyssey. Look up at the night sky and marvel at the celestial wonders above. Seek out knowledge and explore the ever-expanding frontiers of astronomy and physics. Embrace the mysteries that await and let your curiosity guide you.

Remember, the universe is not just a subject of scientific inquiry; it is an invitation to wonder, to dream, and to contemplate our place in the grand cosmic symphony. Whether you are an aspiring astronomer, a curious seeker of knowledge, or simply a lover of the universe's beauty, may this book serve as a catalyst for your own exploration.

Our cosmic odyssey continues, and the universe eagerly awaits our next discovery. So, let us embrace the marvels of the cosmos, with all its beauty, strangeness, and infinite possibilities.

CHEAT SHEET

1. **Big Bang:** The theory that explains the origin of the universe as a colossal explosion from a highly dense and hot state, marking the beginning of space, time, matter, and energy.
2. **Inflation:** A rapid and exponential expansion of the universe in its early stages, believed to account for the observed uniformity and flatness of the cosmos.
3. **String Theory:** A theoretical framework that posits that the fundamental constituents of the universe are tiny, vibrating strings of energy. It aims to unify quantum mechanics and general relativity while suggesting the existence of extra dimensions.
4. **Multiverse:** The hypothetical idea that there exist multiple universes, each with its own set of physical laws and properties, possibly arising from cosmic inflation or other cosmological phenomena.
5. **Dark Matter:** Invisible and unidentified matter that does not interact with light but exerts gravitational influence on galaxies and cosmic structures, accounting for the observed discrepancies between observed and predicted gravitational effects.
6. **Dark Energy:** An enigmatic form of energy that permeates the universe and is responsible for its accelerated expansion. Its exact nature remains unknown.
7. **Quantum Gravity:** The theoretical framework that seeks to describe gravity within the context of quantum mechanics, aiming to reconcile the theory of general relativity with the laws governing subatomic particles.

8. **Anthropic Principle:** The philosophical idea that the fundamental constants and conditions of the universe must be compatible with the existence of intelligent life, explaining the apparent fine-tuning of the cosmos.
9. **Holographic Principle:** The conjecture that the information describing a volume of space can be encoded on its boundary, suggesting a deep connection between gravity and quantum mechanics and the possibility that our three-dimensional reality is a holographic projection from a lower-dimensional surface.
10. **Simulation Hypothesis:** The speculative notion that our reality is a computer-generated simulation, similar to a virtual reality, with an external intelligence or advanced civilization responsible for its creation and maintenance.
11. **Quantum Entanglement:** The phenomenon where two or more particles become inextricably linked, such that the state of one particle instantaneously affects the state of the other(s), even when separated by vast distances.
12. **The Nature of Time:** The philosophical and scientific exploration of the concept of time, including its perception, flow, arrow, and interplay with space and the universe.

ADDITIONAL RESOURCES

1. **Big Bang:**
   - NASA: The Big Bang - https://science.nasa.gov/astrophysics/focus-areas/what-powered-the-big-bang
   - Stanford University: The Big Bang - http://www.preposterousuniverse.com/eternitytohere/bigbang.html

2. **Inflation:**
   - Harvard-Smithsonian Center for Astrophysics: Inflation - https://www.cfa.harvard.edu/~deisenst/acm/acm12_inflation.pdf
   - Berkeley Lab: What is Inflation? - https://www2.lbl.gov/Science-Articles/Archive/Phys-Higgs-Inflation.html

3. **String Theory:**
   - Institute for Advanced Study: What is String Theory? - https://www.ias.edu/ideas/2017/gubser-string-theory
   - Stanford Encyclopedia of Philosophy: String Theory -

https://plato.stanford.edu/entries/qm-string/

4. **Multiverse:**
   - Space.com: Multiverse - https://www.space.com/18811-multiverse-theory.html
   - Scientific American: Parallel Universes - https://www.scientificamerican.com/article/parallel-universes/

5. **Dark Matter:**
   - Fermilab: Dark Matter - https://www.fnal.gov/pub/science/inquiring/questions/dark_matter.html
   - NASA: What is Dark Matter? - https://science.nasa.gov/astrophysics/focus-areas/what-is-dark-energy

6. **Dark Energy:**
   - Berkeley Lab: Dark Energy - https://www2.lbl.gov/Science-Articles/Archive/phys-dark-energy.html
   - NASA: Dark Energy, Dark Matter - https://www.nasa.gov/mission_pages/galex/pia15416.html

7. **Quantum Gravity:**
    - Perimeter Institute: Quantum Gravity - https://perimeterinstitute.ca/research/research-areas/quantum-gravity
    - Stanford Encyclopedia of Philosophy: Quantum Gravity - https://plato.stanford.edu/entries/quantum-gravity/
8. **Anthropic Principle:**
    - Universe Today: Anthropic Principle - https://www.universetoday.com/140073/the-anthropic-principle/
    - Stanford Encyclopedia of Philosophy: Anthropic Principle - https://plato.stanford.edu/entries/anthropic-principle/
9. **Holographic Principle:**
    - Physics World: The Holographic Principle - https://physicsworld.com/a/the-holographic-principle/
    - Perimeter Institute: Holography - https://perimeterinstitute.ca/research/research-areas/holography

10. **Simulation Hypothesis:**
    - Scientific American: The Case for a Creator - https://www.scientificamerican.com/article/the-case-for-a-creator/
    - The Guardian: Are We Living in a Computer Simulation? - https://www.theguardian.com/technology/2016/oct/11/simulated-world-elon-musk-the-matrix

11. **Quantum Entanglement:**
    - Stanford Encyclopedia of Philosophy: Quantum Entanglement - https://plato.stanford.edu/entries/qt-entangle/
    - National Institute of Standards and Technology: Quantum Entanglement - https://www.nist.gov/news-events/news/2018/07/entanglement-now-comes-color

12. **The Nature of Time:**
    - Stanford Encyclopedia of Philosophy: Time - https://plato.stanford.edu/entries/time/

- BBC Science Focus: What is Time? - https://www.sciencefocus.com/science/what-is-time/

Thank you!

Thank you so much for being a part of my literary journey and allowing me to share my thoughts and insights with you.

I strive to create books that provide value and contribute to the knowledge and understanding of various topics. Your feedback is incredibly valuable to me and plays a vital role in shaping my writing style and the subjects I explore.

At OSTRICH, we greatly appreciate the feedback we receive from readers like you. It helps us improve and deliver content that is engaging and meaningful.

We are committed to our mission of creating captivating and informative material, and your input is instrumental in achieving that goal.

I invite you to visit our website at www.ostrichpress.com or check out our books on platforms like Amazon, Kobo, and Google Play. Your thoughts and opinions matter to us, and we would love to hear your feedback on this book or any others you may have read.

Once again, thank you for your support and for being a part of this journey. Your engagement and feedback are truly appreciated.

Warm regards,
Frank Dappah

ABOUT THE AUTHOR

Frank is a charismatic and visionary entrepreneur, accomplished author, and seasoned investor. With a passion for business and a wealth of experience, Frank has written extensively on topics ranging from marketing and social media to entrepreneurship and beyond. His insightful books have garnered praise for their practical advice and actionable strategies.

Living in the vibrant city of Charlotte, North Carolina, Frank thrives on the dynamic energy of the business world. Alongside his partner and wife, Bernice, he has built successful ventures and continues to explore new opportunities in the ever-evolving landscape of entrepreneurship.

Frank's unique perspective and expertise make him a sought-after speaker and advisor, empowering aspiring entrepreneurs to unlock their full potential. With his engaging writing style and knack for simplifying complex concepts, Frank has helped countless readers navigate the challenges and seize the opportunities that come with starting and growing their own businesses.

When he's not immersed in his entrepreneurial endeavors, Frank enjoys spending quality time with his family, exploring the outdoors, and indulging his love for movies, books, and astronomy. His curiosity knows no bounds, and he is always eager to delve into new subjects and expand his knowledge.

Connect with Frank on social media and join him on this exciting journey of innovation, growth, and success.

OSTRICH®

OSTRICH®

OSTRICH PUBLISHERS

Published by Ostrich Publishing Group

Charlotte, North Carolina 28212, U.S.A

First published in the United States of America by Ostrich Publishing Group, an Independent Book publisher.

9798397840149

Copyright © 2022 Ostrich Publishers

www.ostrichpress.com

All rights reserved.

ISBN: 9798397840149

Except in the United States of America, this book is sold subject to the conditions that is shall not, by way of trade or otherwise, be lent, re-sold, hired out, or otherwise circulated without the publisher's prior consent in any form of binding or cover other than in which is it published and without a similar condition including this condition being imposed on the subsequent purchaser.

OUR BIG BEAUTIFULLY STRANGE UNIVERSE

OSTRICH™

FRANK DAPPAH

OSTRICH

Publisher's Disclaimer:

The information contained in this publication is for general informational purposes only. While we have made every effort to provide accurate and up-to-date information, we make no representations or warranties of any kind, express or implied, about the completeness, accuracy, reliability, suitability, or availability with respect to the content contained herein. Any reliance you place on such information is therefore strictly at your own risk.

Content Accuracy:

The content in this publication is based on the knowledge and information available at the time of writing. However, developments in the field may occur after publication, and the publisher cannot guarantee that the information provided will always be complete, accurate, or up-to-date. Readers are advised to consult additional sources and seek professional advice where necessary.

Editorial Responsibility:

The views and opinions expressed by the authors, contributors, and editors of this publication are their own and do not necessarily reflect the views of Ostrich Publishers. The publisher disclaims any liability or responsibility for any errors, omissions, or inaccuracies that may be present in the content.

Legal Compliance:

While every effort has been made to ensure compliance with all applicable laws and regulations, the publisher cannot be held responsible for any legal implications or consequences arising from the use or misuse of the information in this publication. Readers are advised to familiarize themselves with the relevant laws and seek legal counsel if necessary.

Third-Party Content:

This publication may include content from third-party sources, including but not limited to quotes, references, or excerpts. Ostrich Publishers does not endorse or guarantee the accuracy, reliability, or suitability of any third-party content referenced in this publication. Any reliance on such content is at the reader's own discretion and risk.

External Links:

This publication may contain links to external websites or resources. Ostrich Publishers has no control over the nature, content, and availability of those sites or resources. The inclusion of any links does not necessarily imply a recommendation or endorsement by the publisher. Ostrich Publishers shall not be held liable for any damages or losses arising from the use of such external links.

Copyright:

All rights reserved. No part of this publication may be reproduced, distributed, or transmitted in any form or by any means, including photocopying, recording, or other electronic or mechanical methods, without the prior written permission of the publisher, except in the case of brief quotations embodied in critical reviews and certain other noncommercial uses permitted by copyright law.

Contact Information:

For inquiries regarding this publication, please contact:

Ostrich Publishers
Charlotte, NC
U.S.A
Email: admin@ostrichpress.com
Website: www.ostrichpress.com

Disclaimer Updates:

Ostrich Publishers reserves the right to amend or update this disclaimer at any time without prior notice. It is the responsibility of the readers to regularly review this disclaimer for any changes.

Last Updated:
May 202

OUR BIG BEAUTIFULLY STRANGE UNIVERSE

"A Non-Astronomer's Take on the Big Bang, Inflation, and the Wondrous Quirks of Our Universe"

FRANK DAPPAH

OSTRICH

www.ingramcontent.com/pod-product-compliance
Lightning Source LLC
Chambersburg PA
CBHW020433220526
45464CB00002B/683